Housing 121

你说哪种语言？

Which Languages Do You Speak?

Gunter Pauli

［比］冈特·鲍利　著

［哥伦］凯瑟琳娜·巴赫　绘

唐继荣　译

上海遠東出版社

特别感谢以下热心人士对童书工作的支持：

匡志强　宋小华　解　东　厉　云　李　婧　庞英元
李　阳　梁婧婧　刘　丹　冯家宝　熊彩虹　罗淑怡
旷　婉　王靖雯　廖清州　王怡然　王　征　邵　杰
陈强林　陈　果　罗　佳　闫　艳　谢　露　张修博
陈梦竹　刘　灿　李　丹　郭　雯　戴　虹

目录

Contents

一只北美山雀在树上遇到一只松鼠。松鼠正努力干活，采集坚果好过冬。

“打扰一下，请问你说什么语言？”北美山雀问松鼠。

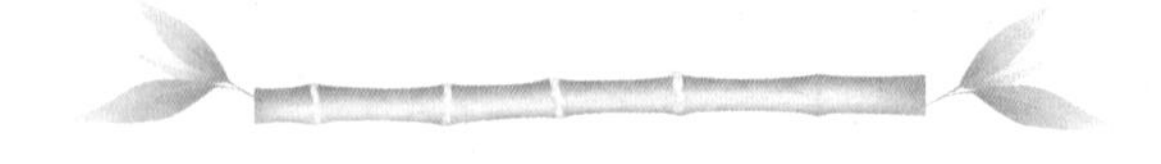

A chickadee meets a squirrel in a tree. The squirrel is hard at work, collecting nuts for the winter.

"Excuse me, what language do you speak?" the chickadee asks the squirrel.

一只北美山雀在树上遇到一只松鼠。

A chickadee meets a squirrel in a tree.

我能说两种语言。

I can speak two languages.

“唔，我能说两种语言。第一种，我当然会说松鼠语；第二种，我的鸟语也说得非常流畅。我还一直在研究你们北美山雀的方言，你们说什么我都能听懂。”松鼠回答。

“真厉害！”北美山雀说，“我必须承认，我不擅长语言。不过既然你会说我的语言，我为什么还要学习你的语言呢？”

“Well, I can speak two languages. First, of course, I speak Squirrelese, and secondly, I am pretty fluent in Birdese. I have also been studying your chickadee dialect – and I am getting quite good at understanding all you say,” responds the squirrel.

“Very impressive!” Chickadee says. “I must admit I am not good at languages. Anyway, if you can speak mine, why should I learn yours?”

“我发现，每当你学习另一种语言，你就会了解一个全新的世界。”

“我猜想应该还有更多的收获。你每学一种语言，你就会更多地使用大脑，这让你更聪明！”

“I find every time you learn another language, you get to know a whole other world.”

“I imagine there is more to it than just that. With every language you learn, you use more of your brain, and that makes you smarter!”

……你就会了解一个全新的世界。

... you get to know a whole other world.

……用7种不同的语言……

... in seven different languages ...

“哦！我知道有人能用7种不同的语言交谈、发火和恋爱。”松鼠说道，咧嘴笑了笑。

“真的？我很好奇，那人的大脑得有多大？这样的语言学习能力肯定不适合我这样的小脑袋，当然也不适合你的脑袋。”

“Oh, I know of someone who can converse, get mad and fall in love – in seven different languages,” Squirrel shares, with a big grin.
“Really? I wonder what brain size that person has? It will definitely not fit into my little skull – and not into yours either!”

“北美山雀，你这么聪明，应该学习另一种语言。这会让你终身受益，而且毫无疑问，你将结识很多新朋友。”松鼠说道。

“那么，你说说哪一种语言是我最需要学的，会让我的生活更有趣呢？”北美山雀问。

“You are so bright, Chickadee, you should learn another language. It may serve you well in life, and you will no doubt make many new friends,” Squirrel says.

“So, which language do you suggest I need most, one that would make my life more interesting?” Chickadee asks.

……结识很多新朋友……

... make many new friends ...

……猛禽语，苍鹰的语言……

... Raptorese, spoken by the goshawks ...

“我毫不怀疑，学习猛禽语绝对能帮到你，就是你的捕食者——苍鹰的语言。”松鼠提议道。

“这可太难了。”北美山雀迅速回答，“我一听到它们的叫声，就知道小命快要不保！这是我能从它们的尖叫声中得到的全部了。我宁可学习猫头鹰语。”

“I have no doubt that it will serve you well to learn Raptorese, the language spoken by your predators, the goshawks,” Squirrel proposes.

“But that will be too difficult,” Chickadee quickly replies. “When I hear their calls, I know my life is in danger! That is all I can make out from their shrieks. I would rather learn Owlese.”

“猫头鹰语？为什么是猫头鹰语？这和你们北美山雀语不是差不多？你们只不过在每个句子末尾添加一堆‘嘚’的音节。你就不能想出一种更具挑战性的语言去学吗？”

“我哪有时间去学那些？松鼠，你知道的，仅仅是搜集每天需要的食物已经让我非常忙碌了。或许，我有点懒……”北美山雀坦白道。

“Owlese? Why Owlese? Is it not pretty much the same as Chickadese, where you just add a bunch of “dee” syllables to the end of every sentence? Can’t you think of a more challenging language to learn?”

“Where would I find the time to do that? You know, Squirrel, just gathering my daily food already keeps me very busy. And perhaps I am a bit lazy ...” Chickadee confesses.

猫头鹰语……

Owlese ...

……城市中这么多的噪音……

… all the noise from the city …

“瞧，在森林中，我们不得不在吃东西的时间和提防危险的时间上找到平衡。如果我们都能听懂其他语言，就可以分担警戒的责任。这将给同伴更多时间去找寻食物。”

“你说得很对。”北美山雀同意道，“但是，城市中的人们制造出这么多的噪音，我很难听清其他同伴在说什么！从咆哮的汽车发动机到加速的摩托车，更别提持续不断的喇叭声了。这些声音震耳欲聋，我几乎听不到朋友们的报警声，更别提听懂了。”

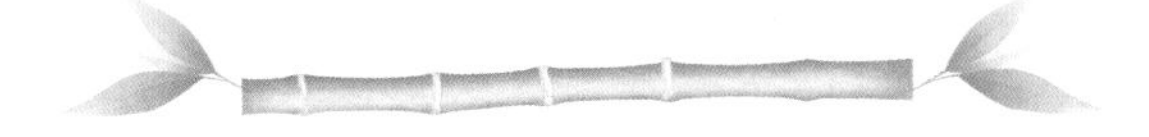

“Look, in the forest we all have to find the balance between time for eating and time for being on the lookout for danger. If we understand the languages of others, we can all share the responsibility of being vigilant. That leaves more time for everyone to find food.”

“You are so right,” Chickadee agrees. “But it is difficult to hear the others with all this noise that people in the city make! From roaring car engines and accelerating motorbikes – not to mention this constant honking. It is quite deafening! I can hardly hear my friends’ warning calls, let alone understand them.”

“是啊。如果人们能学习我们的语言该多好，他们就会理解为什么我们要抱怨他们的噪音强度了！”松鼠说道。

“嗯，好消息！我相信在我们说话时，这种情况正在发生。”北美山雀说着，露出了大大的笑容。

……这仅仅是开始！……

“Yes, if only people could learn our languages, they would understand the reasons we complain about their noise levels!” Squirrel says.

“Well, good news! I believe this is happening as we speak,” Chickadee says with a big smile.

... AND IT HAS ONLY JUST BEGUN! ...

……这仅仅是开始！……

... AND IT HAS ONLY JUST BEGUN! ...

Did You Know?

你知道吗？

Animal communication can be auditory, visual and also chemical. Certain animals can even tell which member of their group is making an alarm call, and if that one is reliable.

动物可以采取声音、视觉和化学方式进行交流。某些动物甚至能区分它们群体中的哪一个成员正在发出报警声，以及该报警是否可信。

Animals use different calls for warning against specific animals. This demonstrates mastery of semantics by non-humans. Vervet monkeys use a different warning call for an eagle, a leopard and a snake. Their calls can even indicate the likely direction of the approaching danger.

对特定动物，动物用不同的叫声来发出警报，这证明了非人类的动物对语义的掌握。长尾猴对老鹰、豹子和蛇发出不同的报警声，它们的叫声甚至能指示正在接近的危险的可能方位。

Some alarm calls are deceptive, for example a male giving a false predator alarm call when a female shows interest in another male. Some alarm calls chase away fellow birds, allowing the caller to continue feeding undisturbed.

一些报警声是有欺骗性的。例如，当一只雄性个体发现一只雌性个体表现出对其他雄性感兴趣时会给出虚假的捕食者到来的报警声，以干扰它们的交往。一些报警声能赶走其他同类，让发声个体能不受干扰地继续觅食。

Animals and plants use chemicals (called pheromones) to broadcast an alarm signal. When a catfish is hurt, it releases a pheromone that will cause fish nearby to form dense schools near the bottom of the river or pond.

动植物利用被称为信息素的化学物质来散播报警信号。当一条鲶鱼受伤时，它释放出的信息素会使附近的鱼在河流或池塘底部形成密集的鱼群。

Being multilingual can improve attention and memory, and provide a "cognitive reserve" that delays the onset of dementia. Speaking four or more languages fluently, offers an extra nine years of healthy cognition.

掌握多种语言能提高注意力和记忆力，其提供的“认知储备”可以推迟痴呆的发病。能流利地使用 4 种以上语言的人的健康认知可以延长 9 年。

Learning to speak another language increases your capacity to adapt: for instance French is more romantic and Italian is more passionate, allowing the adoption of different behaviour according to the language spoken.

学会说另一种语言能增强你的适应能力，让你可以根据所说的语言来调整不同的行为，例如，法语更浪漫，意大利语更富有激情。

Different languages evoke different memories of life. Adopting a new cultural identity and building up associated memories, build neural barriers between languages – avoiding confusion with the mother tongue.

不同的语言唤起不同的生活记忆。采用一种新的文化身份认同和构建关联记忆会在语言之间建立起神经屏障，以避免与母语混淆。

Facial expressions are crucial in making sound. Pouted lips make you sound more French. So learning a new language opens up a new world.

在发声时面部表情非常重要。撅起的嘴唇使你的声音听起来更像法语。所以学习一种新的语言就开启了一扇通往新世界的大门。

Would you like to speak many languages?

你想掌握多种语言吗?

Do you believe that birds can "talk" to each other? And that birds can communicate with plants?

你相信鸟类之间能互相"交谈"吗?你相信鸟类能与植物交流吗?

Is a person who can speak many languages a smart person?

一个能说多种语言的人很聪明吗?

Will you behave like a Frenchmen when you speak good French, or like a Chinese person when you master the Chinese language?

当你法语说得很流利时,你的举止会像法国人吗?或当你精通中国话时,你的举止会像中国人吗?

Do It Yourself!

自己动手！

Time to learn a few keywords in ten different languages. First determine which languages, other than your native tongue, you would like to use. Now pick three words or phrases that you would like to learn, for example: Thank you, Good morning and My pleasure. Have these translated into all ten languages. Make sure that you get the phonetic writing, or even an audio version, so that you can practice the pronunciation. Start by learning these words or phrases in three, then in five, and finally in all ten languages. When learning these words or phrases off by heart, think of them as music and get the intonation right. If you find that learning them in ten languages is easy, then be bold enough to learn them in twenty!

是时候学习用 10 种不同语言说一些关键词了。首先确定除母语外，你想用哪些语言。挑选 3 个你想要学习的词或短语，如“谢谢你”“早安”和“乐意效劳”。把这些词语翻译成这 10 种语言。确保得到音标文本甚至音频版，这样你就可以练习发音。先学习用 3 种语言说这些词语，然后拓展到 5 种，最后是所有 10 种语言。用心背熟这些词语，把它们当作音乐，并注意语调正确。如果你发现学会用 10 种语言说这些词语很容易，就可以大胆地学用 20 种语言说这些词语。

学科知识

Academic Knowledge

生物学	灵长类中的原始语言；北美山雀大脑容量小，但包含旧信息的神经元（神经细胞）不断地被储存新信息的新神经细胞替代；北美山雀是囤积者，为未来储存食物；北美山雀是非迁徙性鸟类，在冬季需要比夏季多达10倍的食物；北美山雀是最重要的害虫消灭者之一。
化　学	信息素在动物交流中的作用。
物　理	通过只有几十亿分之几米至微米尺寸的微小颗粒进行化学通讯；噪音污染。
工程学	基于人工智能的自动翻译。
经济学	翻译业务正在快速增长；全世界教育花费最高的是英语学习。
伦理学	当一个国家的人把不同的语言作为母语时，怎么能强制全国使用单一语言呢?
历　史	巴别塔（古代巴比伦未建成的通天塔）的故事；世界语作为通用性世界语言的设计和消亡。
地　理	印度次大陆拥有约600种语言。
数　学	北美山雀15种不同叫声语序的发展。
生活方式	人们渴望说他们当地的方言；英语是如何将它自身作为一门通用世界语强加于人，并因此导致文化多样性的丧失。
社会学	在一些社群中，人们如何学会把不同的语言当作日常生活的组成部分，例如丹麦人、瑞士人、印度人和南非人；对那些拥有许多不同语言的国家，怎样寻求一门通用语，如印度的印地语、印度尼西亚的印尼语和南非的英语；当一种语言有很多方言时，那些拥有这类语言的国家怎样寻求唯一独特的方言通用语发音，如德国人的标准高地德语与荷兰的标准荷兰语。
心理学	当叫声会吸引捕食者注意、导致发声者付出生命代价时，为什么动物还要发出报警声？把你的敌人的语言当作一种防御策略来学。
系统论	人们对生物多样性的丧失很关心，但还存在文化多样性的丧失；随着一些文化被更主流的文化吸收，数百种语言正在消失。

情感智慧

Emotional Intelligence

北美山雀

北美山雀对松鼠的交流能力很好奇。它对此印象深刻，承认自己无法掌握更多语言。不过，山雀找到了一种让它继续悠闲的逻辑：既然松鼠学习了北美山雀的语言，为什么自己还要学习松鼠的语言？山雀认识到一个事实，了解更多语言可以更好地利用大脑，它开玩笑说某个说7种语言的人有比它或松鼠更庞大的大脑。北美山雀被劝诱着去学习，它先征求松鼠的建议，但随后表示还没准备好冒险学习一门新语言，认为这太困难。当受到松鼠的挑战时，这只鸟儿就寻找借口，宣称周围噪音太多，所以它不能集中精力学习。

松　鼠

松鼠对使用不同的语言有自信，它告诉山雀自己理解北美山雀方言已经完全没有障碍。它分享了一种人生哲理：学习一种新的语言就是为了了解一个新的世界。松鼠鼓励北美山雀去学另一种语言；它称赞山雀的智慧，提示说这将帮助它在生活中做得更好。当那只山雀寻求建议时，松鼠提议它学习其捕食者的语言，希望学习的结果能激励这只鸟儿。当山雀表示它愿意学习一种与其自身语言相似、容易的语言时，松鼠迅速判定这个主意是噱头。松鼠敦促山雀认真对待生活，并坚持学习一门新的语言是积极生活的一部分。

艺术

The Arts

来看看如何用不同的语言表达“爱爸爸爱妈妈”，比如用德语、英语或西班牙语。听听语调和发音。你认为哪一种语言的表达是孩子们最乐意展现给他们父母的？用不同语言来表达你的情感的艺术能让你体会到情感如何有效地交流；不仅知道用哪些单词，还懂得这些表达在不同语言之间的不同，甚至文化上的差异。这真是一门艺术！

思维拓展

Systems: Making the Connections

在全球化背景下，我们越发需要理解他人并跨越文化和地域去与他人交流的能力。然而，不是每个人都有天赋或意愿去研究其他语言，并且打破语言藩篱去沟通。我们所有人都拥有学习不止一种语言的能力，每次我们大胆地开口说一种外语时，就会有一个全新的世界向我们敞开。语言隔绝不仅阻止我们更多地向他人学习，而且妨碍我们积极地利用大脑的全部能力。经证明，学习4种以上语言的人患痴呆和（或）记忆丧失的风险会降低；同时有更多的证据表明，动物、植物甚至真菌，不仅有它们自己内部交流能力，还能理解甚至模拟其他物种的语言。通过掌握其他语言来弥合分歧，这种力量是有据可查的，该是时候认识到，如果不努力，我们将在一个互相连接和依存的世界里自我孤立。直到最近，倾听、理解和与他人交流的能力都仅限于人群和人类文化。而我们正缓慢但稳步地揭示出，交流是多功能、多感官、多模式和多物种的。这为我们提供了大量的机会去了解更多其他有知觉的物种，而这些构成我们生态系统的一部分。我们正在进入一个新的时代，通过互联网以及许多不同的应用程序和软件系统，拓展我们的交流。通过新技术实现的信息传递爆发已经被广泛描述。一种生物与另一种生物之间、一个社会与另一个社会之间的交流时机为我们提供了一个向完整的新世界开放的机会，而这个新世界有对传统、文化和社会更好和更深的理解。这无疑将有助于我们创造更美好的世界。

动手能力

Capacity to Implement

你的朋友和家人能说多少种语言？他们都愿意去学的语言有多少种？他们希望怎样学习这些语言？他们喜欢如何使用新近获得的语言技能？听了这则故事后，是否有人对学习鸟类语言和理解鸟类叫声有兴趣？找到这个领域的专家，组织一次课程，为人们提供理解不同鸟类叫声甚至是松鼠和其他动物的语言的独特机会。在帮助他人接受如此独特的语言培训之前，你自己也要学会认识和理解这些叫声。

故事灵感来自

This Fable Is Inspired by

埃里克·格林

Erick Greene

埃里克·格林是美国米苏拉蒙大拿大学的生物学教授。他先是从达尔豪斯大学毕业，获得生物学与数学学位，然后于1989年获得生态学与进化生物学的博士学位。他是动物通讯领域的专家，研究对象从鸟类和蜘蛛到苍蝇和毛虫。他甚至还研究了植物通讯。他分析鸟类的报警鸣叫声，发现北美山雀是极佳的鸟类观察者。他还研究一波叫声怎样以令人惊异的速度席卷森林，以及发出与鸟类相似声音的红松鼠是如何加入进来的。格林教授还研究鸟类如何在它们的第一个繁殖季学会唱歌。通过研究，他发现鹗（又称鱼鹰）这种集大群筑巢的猛禽会建立一个信息中心，它们在这个中心与其他鹗个体交换有关鱼群分布的信息。他是最早揭开这个新的交流世界的科学家之一。

图书在版编目(CIP)数据

冈特生态童书.第四辑：修订版：全36册：汉英对照 /
(比)冈特·鲍利著;(哥伦)凯瑟琳娜·巴赫绘；
何家振等译.—上海：上海远东出版社,2023
书名原文：Gunter's Fables
ISBN 978-7-5476-1931-5

Ⅰ.①冈… Ⅱ.①冈… ②凯… ③何… Ⅲ.①生态环
境-环境保护-儿童读物—汉、英 Ⅳ.①X171.1-49

中国国家版本馆CIP数据核字(2023)第120983号

著作权合同登记号图字09-2023-0612号

策　　划　张　蓉
责任编辑　张君钦
封面设计　魏　来李　廉

冈特生态童书
你说哪种语言?
[比]冈特·鲍利　著
[哥伦]凯瑟琳娜·巴赫　绘
唐继荣　译